Ardeshir Hezarkhani
Seyyed Saeed Ghannadpour

Geochemical Behavior Investigation Based on K-Means Clustering

AF153289

Ardeshir Hezarkhani
Seyyed Saeed Ghannadpour

Geochemical Behavior Investigation Based on K-Means Clustering

Basics, Concepts and Case Study

LAP LAMBERT Academic Publishing

Impressum / Imprint
Bibliografische Information der Deutschen Nationalbibliothek: Die Deutsche Nationalbibliothek verzeichnet diese Publikation in der Deutschen Nationalbibliografie; detaillierte bibliografische Daten sind im Internet über http://dnb.d-nb.de abrufbar.
Alle in diesem Buch genannten Marken und Produktnamen unterliegen warenzeichen-, marken- oder patentrechtlichem Schutz bzw. sind Warenzeichen oder eingetragene Warenzeichen der jeweiligen Inhaber. Die Wiedergabe von Marken, Produktnamen, Gebrauchsnamen, Handelsnamen, Warenbezeichnungen u.s.w. in diesem Werk berechtigt auch ohne besondere Kennzeichnung nicht zu der Annahme, dass solche Namen im Sinne der Warenzeichen- und Markenschutzgesetzgebung als frei zu betrachten wären und daher von jedermann benutzt werden dürften.

Bibliographic information published by the Deutsche Nationalbibliothek: The Deutsche Nationalbibliothek lists this publication in the Deutsche Nationalbibliografie; detailed bibliographic data are available in the Internet at http://dnb.d-nb.de.
Any brand names and product names mentioned in this book are subject to trademark, brand or patent protection and are trademarks or registered trademarks of their respective holders. The use of brand names, product names, common names, trade names, product descriptions etc. even without a particular marking in this work is in no way to be construed to mean that such names may be regarded as unrestricted in respect of trademark and brand protection legislation and could thus be used by anyone.

Coverbild / Cover image: www.ingimage.com

Verlag / Publisher:
LAP LAMBERT Academic Publishing
ist ein Imprint der / is a trademark of
OmniScriptum GmbH & Co. KG
Heinrich-Böcking-Str. 6-8, 66121 Saarbrücken, Deutschland / Germany
Email: info@lap-publishing.com

Herstellung: siehe letzte Seite /
Printed at: see last page
ISBN: 978-3-659-81121-0

Zugl. / Approved by: Tehran, Amirkabir University of Technology (Tehran Polytechnic), Diss., 2013.

Geochemical behavior investigation based on K-Means Clustering

Basics, Concepts and Case Study

Ardeshir Hezarkhani [1*]; Seyyed Saeed Ghannadpour [1]

[1] Department of Mining & Metallurgical Engineering, Amirkabir University of Technology

(Tehran Polytechnic), Tehran, Iran

---

[*] "Ardeshir Hezarkhani". Tel.: +98-912-187-2494
E-mail address: ardehez@aut.ac.ir

Abstract

A well-known algorithm of clustering is K-means by which the data are divided into K classes based upon a distance criterion. In this study, initially, the K-means method applies for data classification derived from exploration boreholes in the Parkam deposit. Subsequently, it is applied in order to estimate the Cu grade in case of surface data set. The optimum K has been calculated and then the data have been clustered and the relative geochemical behavioral characteristics analyzed. The criterion used for determining the optimum K ranged in number of classes from K=3 to K=10 and afterwards, we analyzed derived classifications in order to choose the optimum K. The results derived based on the case of borehole data set showed that class numbers of K=3 in the case of Cu and Mo, K=4 in the case of Cu and Pb, and K=3 in the case of Cu and Zn were optimized class numbers. After clustering, the increasing Cu grade values resulted in a significant increase in Mo grades, a significant decrease in Pb grades followed by an increase, and the Zn grades varying comparable to Pb. With regards to the relationships between these elements it can be concluded that the meteoric waters promoted the mobilization of Pb and Zn from the potassic zone to the phyllic but the meteoric waters were not effective enough to cause the mobilization of Cu, and this element together with Mo remained immobile. Moreover, it sees that the class numbers of K=5 in the case of Cu and Mo for surface data set, considering the latitude and longitude of the samples was optimized in class numbers. Finally, equation of Cu grade estimation, according to the mentioned five cluster centers are presented with the well correlation coefficient.

Keywords

K-Means, Clustering, Geochemical behavior, Grade estimation, Parkam.

Table of contents

1- Introduction

Considering the Miduk and Parkam areas located within the Kerman Copper Belt (with a length of 450 and the width of 80 kilometers) and the presence of copper-rich resources in the region, the necessity of determination the geochemical behavior of Cu, Mo, Pb, Zn and Fe and also the mentioned elements grade estimation within these deposits is obvious. The geochemical data were obtained from samples from exploration drilling (coring) and surface sampling in the Parkam exploration area and were treated by initial statistical characteristics (Ghannadpour and Hezarkhani 2012; Ghannadpour, 2013; Ghannadpour et al. 2014, Ghannadpour et al. 2015a; Ghannadpour and Hezarkhani, 2015a; Hezarkhani and Ghannadpour, 2015). In addition to Cu and Mo which are major ore-forming elements in porphyry systems, Pb and Zn have also been examined as these have significant impact on estimating the extent to which a primary geochemical halo is expanded (Gent et al. 2011). In some cases, Pb and Zn are observed to show concentrations of ore grade (Jébrak 2006). Hence to make a more effective estimation of geochemical halo boundaries in porphyry systems, relative behavioral characteristics of these elements need to be investigated. Thus in the present research, methods of data mining sciences were utilized for performing the analysis.

One of the most important approaches in data mining science concerns with analysis of large amounts of data and samples of different properties through clustering, using different methods and techniques such as hierarchical K-means method, density-based methods, Kohonen's method (e.g., Devijver and Kittler 1982; Anderberg 1973). Clustering analysis involves the collection M including m specimens in the form of $(x_1, x_2, ..., x_3)$, to each of which a vector in collection M is attributed and the vector represents different characteristics of the specimen (Nelson et al. 2012). Assume that these specimens are to be classified into K classes or groups. For this purpose, some basic justifiability criteria must be satisfied (Jain 2010; Pelleg and Moore 1999). The assumptions are as follows:

(1) $$C_i \neq \varnothing, \quad for \; i = 1, ..., K$$

(2) $$C_i \cap C_j = \varnothing, \quad for \; i \neq j$$

(3) $$\bigcup_{i=1}^{K} C_i = M$$

According to the first assumption, no one of K class must be null. The second assumption states that the collections must not overlap and the third assumption states that no specimen must be out of collection.

K-means has been one of the well-known and simple methods in which the sum of Euclidean distances of specimens to the center of associated collection must be minimized. Relative geochemical behavioral characteristics have been analyzed through different approaches (Menard 1995; Xu et al. 2012; Tarkian and Stribrny 1999). There are numerous studies utilizing clustering methods especially K-means for classifying the data populations associated with

6

geological variables and often of a geochemical nature. The following could be indicated as examples: classifying land features (Yang et al. 2012), classifying the effect of vegetation on water healthiness recovery in Mediterranean forests (Mora et al. 2012), planning the identification of geochemical patterns in mining districts applying K-means method (Meshkani et al. 2011), organic carbon prediction by intelligent systems comparing K-means and intelligent methods (Sfidari, et al. 2012), and determining the gas emission effect in urban environments applying clustering analysis by K-means method (Wegner et al. 2012). Moreover, there have been researches applying K-means with advanced innovative algorithms which yielded favorable results (Krishna and Narasimha Murty 1999; Cheung 2003; Murthy and Chowdhury 1996). Yaghini et al (2008) devised a combined clustering method (GKA) engaging both genetic and K-means algorithms. The latter method utilizes K-means operator which is designed based upon K-means algorithm.

Investigation of Pb geochemical behavior with respect to those of Fe and Zn in Parkam porphyry systems based on K-means clustering method, has been reported in Ghannadpour et al (2013). In this research, the analysis of Cu, Mo, Pb and Zn relative geochemical characteristics will be evaluated in the case of borehole data set. And also, Cu grade will be estimated considering the latitude, longitude and Mo grade of samples in the case of surface data set. Due to the number of available data, data-driven methods such as neural network, fuzzy clustering and so on cannot be used. Data-driven methods such as mentioned ones need too much data for training and their training will be the main problem while we are facing low number of available data (Lawrence et al. 1996; Tahmasebi and Hezarkhani 2012); however K-means method does not have any problem in this case and could perform classification and monitoring the results well. It could be referred to the study by Ghannadpour (2013) that in the case of surficial samples of Parkam district, the K-means method is more efficient than the neural network method (due to the low number of data).

2- K-Means Algorithm

The K-means is also one of the well-known viewpoints in data mining clustering method which starts with a certain number of the collections (K) and cluster specimens into these K group so that the assumptions 4 and 5 are satisfied (Saha and Bandyopadhyay 2012). The criterion based upon which specimens are attributed to collections is the minimum Euclidean distance between each specimen and the central point (representative point) of each group. The most important stages of K-means algorithm have been defined by Bock (Bock 2008) and Jain (Jain 2010) as follows:

1. Introducing K class or group as $(C_1, C_2, ..., C_k)$ in order for clustering m specimens from the collection M.

2. Calculating the vector zj (based on equation 4) which is representative of each class Cj.

(4)
$$z_j = \frac{\sum_{x \in C_j} x}{\#C_j} \quad for\ j = 1,2...,K$$

The parameter x in equation 4 stands for the vector of a specimen which is a member of Cj and #Cj stands for the number of specimens that are members of Cj (this equation is used for calculating the class central points while solving the algorithm begins with random number of specimens (k) and attributes them to the center of each group.

3. Calculating of the objective function derived from classification (C_1, C_2, ..., C_k) based on equation (5) which calculated the sum of distances between specimens and central point.

(5)
$$f(C_1, C_2, ..., C_k) = \sum_{j-1}^{k} \sum_{x \in C_j} |x - z_j|^2$$

4. Minimizing objective function (f) and determining appropriate classification for collection M with K number of classes. Actually for each certain number of collections (k), there could be different collections with associated value objective functions (f). Therefore, minimizing objective function (f) means selecting minimum value of f among possible n state of classification for a constant k (Ghannadpour and Hezarkhani, 2015b). The operation procedure is briefly illustrated in Fig. 1.

1. *Let $C_1, C_2, \ldots, C_k$ be a set of k clusters of M.*
2. *Let $Z_j = (\sum_{x \in C_j} x)/\#C_j$ for $j = 1, 2, \ldots, k$, where x is a pattern vector in C_j and $\# C_j$ represents the number of points in C_j.*
3. *Let $f(C_1, C_2, \ldots, C_k) = \sum_{j=1}^{k} \sum_{x \in C_j} \|x - z_j\|^2$. we shall refer to $f(C_1, C_2, \ldots, C_k)$ as the objective function of the clustering $C_1, C_2, \ldots, C_k$.*
4. *Minimize $f(C_1, C_2, \ldots, C_k)$ over all $C_1, C_2, \ldots, C_k$ satisfying P_1, P_2 and P_3 stated in Section 1.*

Fig. 1 The performance of K-Means Algorithm (Ghannadpour et al. 2013).

3- Regional Geology

The small Parkam stock is part of the Sahand-Bazman igneous and metallogenic zone, a deeply eroded Tertiary volcanic field, approximately 100 × 1700 km^2 in extent (from Turkey to Baluchistan in southern Iran), which consists mainly of rhyolite and andesite, with numerous felsic intrusions (Fig.2). The litho-tectonic units shown in Figure 2 formed as a result of the opening and closing of the Paleo–Tethys and Neo–Tethys oceanic basins, due to subduction and terminal collision and transpression events. To the north, Iran collided with Turkmenistan or Turan plate (Eurasia) in the Late Triassic–Early Jurassic period (Berberian and King, 1981). In the south, Iran experienced subduction and collided with the Arabian Plate in the Late Cretaceous (Takin, 1972; Stöcklin, 1974, 1977; Hallam, 1976; Welland and Mitchell, 1977; Adamia et al. 1980; Berberian and King, 1981). Arc magmatism continued through the Miocene and Late Neogene (Alavi, 1994; Walker and Jackson, 2002; Shahabpour, 2005), which resulted in extensive alkaline and calc-alkaline volcanic and plutonic igneous activities

(Etminan, 1978; Shahabpour, 1982; Berberian, 1983; Hezarkhani, 2006a, 2006b), including the Miocene intrusion of a porphyritic calc-alkaline stock at Parkam (Shahabpour, 1982).

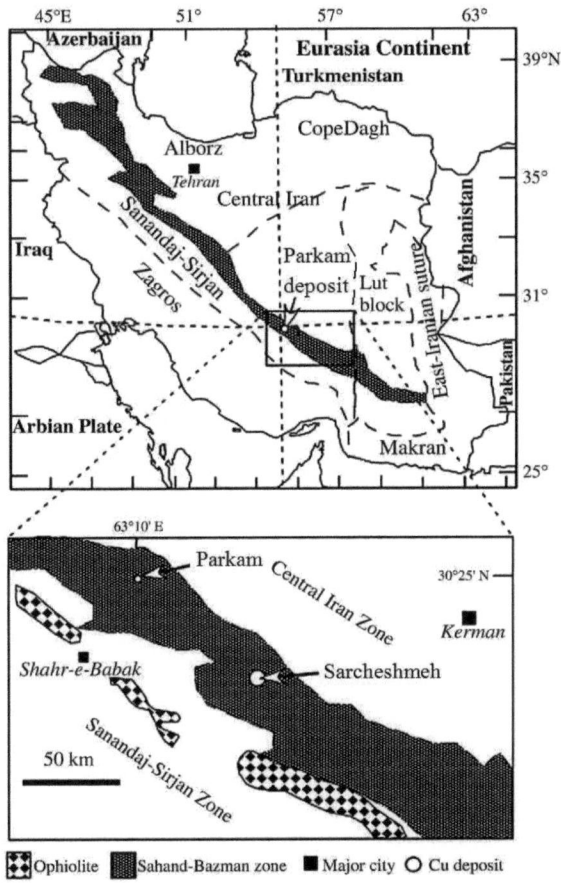

Fig. 2 Map showing major litho-tectonic structural zones of Iran. Enlarged part shows location of the Parkam porphyry Cu deposit, the giant Sarcheshmeh porphyry Cu–Mo deposit and the city of Shahr-e-Babak. Stippled zone is Sahand–Bazman (modified after Berberian and King 1981).

The Sahand-Bazman zone (Fig. 2) encompasses the southeastern part of the Central Iranian Cenozoic volcanic arc known as the Kerman Copper Belt, which represents a length of 450 kilometers and a width of 80 kilometers, and hosts the well-known Iranian middle/late Miocene porphyry copper ore deposits in the Kerman Province such as Miduk (diorite-type) and Sar-Cheshmeh (granodiorite-type), as well as several other porphyry copper systems including Parkam. A faulted and slightly-folded Upper Eocene-Oligocene complex of poorly-bedded lavas, volcaniclastic rocks, and subordinate sedimentary deposits hosts the Neogene intrusions and their associated copper mineralization in the Kerman Copper Belt.

4- Geology of Parkam district

Porphyry system of Parkam (Sarah) with the coordinates of 55° 08′ 54″ E and 30° 26′ 24″ N (Fig. 3a) at a distance of 50 kilometers from north of Shahr-e Babak is located on the metallogenic belt of Kerman. Kerman belt which forms southern section of metallogenic country of Urumieh-Dokhtar (Sahand-Bazman), is the most significant porphyry copper belt in Iran (Hezarkhani, 2006a, 2006b) (Fig. 2). In this belt with a length of 450 kilometers and a width of 80 kilometers, there are more than 200 ore deposits and indexes known to exist that some of them like Parkam ore deposit are from porphyry type.

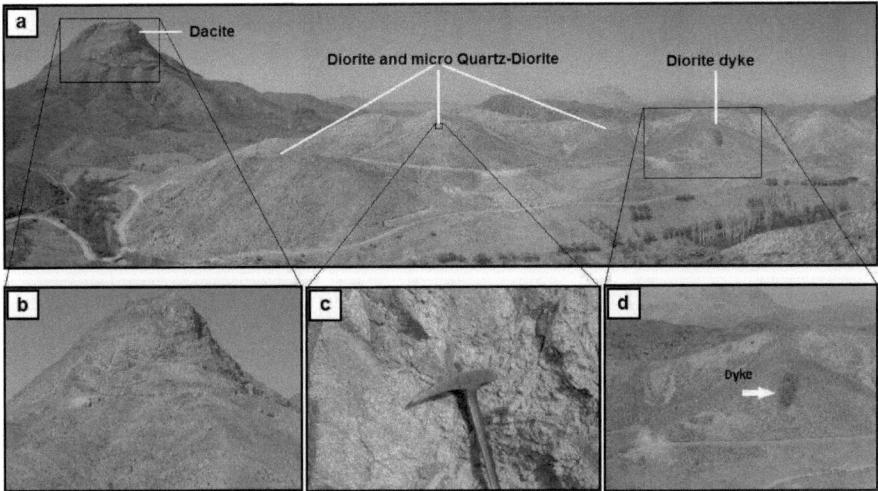

Fig. 3 The Parkam exploration district. a) Panorama of the study area (view to the north). b) Enlarged image of a dacite dome; c) Enlarged image of the diorite to micro quartz-diorite subvolcanic body (the Parkam porphyry); d) Enlarged image of a diorite dyke.

The geologic map of the Parkam exploration district was prepared based on detailed field studies (Fig. 4). According to observations, a set of volcanic rocks accompanied by recent unconsolidated sediments comprise the host to the subvolcanic body and associated porphyry copper mineralization in the Parkam exploration district. The volcanics vary from mafic to felsic, but are mostly intermediate in composition and include andesitic volcanic rocks, basaltic to andesitic lavas, and andesitic pyroclastic rocks widespread in the area (Fig. 5a; Fig. 4). These lithological units are intruded by diorite dykes (Fig. 3d) and subvolcanic bodies of diorite and micro quartz-diorite (Fig. 3c; Fig. 4) which are associated with alteration system (see the next section; see also Tangestani and Moore, 2001). The latest stage of magmatism was accompanied by the emplacement of the dacite dome in the northwestern part of the district (Fig. 3b).

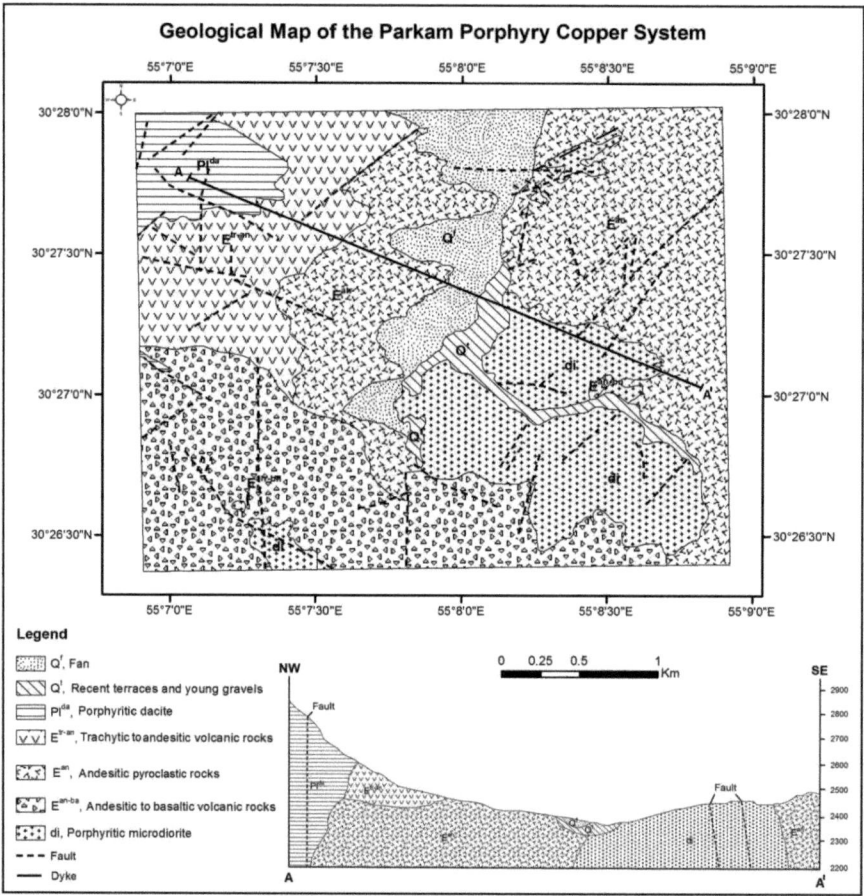

Fig. 4. Geologic map of the Parkam exploration district with a cross-section describing the vertical and lateral distribution of different lithological units.

Fig. 5. Pyroclastic rocks of the Parkam area. a) Andesitic pyroclastic rocks which are widespread in the study area and constitute the main host lithology to the porphyry system; b) Epidote in the fractures of andesitic pyroclastic rocks indicating propylitic alteration.

4-1- Alterations and mineralization

Hydrothermal alteration is widespread in the Parkam area and includes propylitic, phyllic, argillic, and potassic types in order of extent. The phyllic alteration constitutes the central zone of the alteration system and is associated with the potassic alteration in the east and west, as well as with the argillic alteration toward the south. The propylitic alteration surrounds the other types of alteration in the study area (Fig. 6) (Ghannadpour et al. 2015c). The presence of the potassic alteration in the peripheral zone of the porphyry system (in addition to the central zone) is a characteristic of the Parkam exploration district (Fig. 4; Fig. 6). Expect for the propylitic alteration which extends mostly over the host volcanic rocks (Fig. 5b), the other types of alteration are associated mainly with the dioritic subvolcanic body (Fig. 4; Fig. 6; Fig. 7).

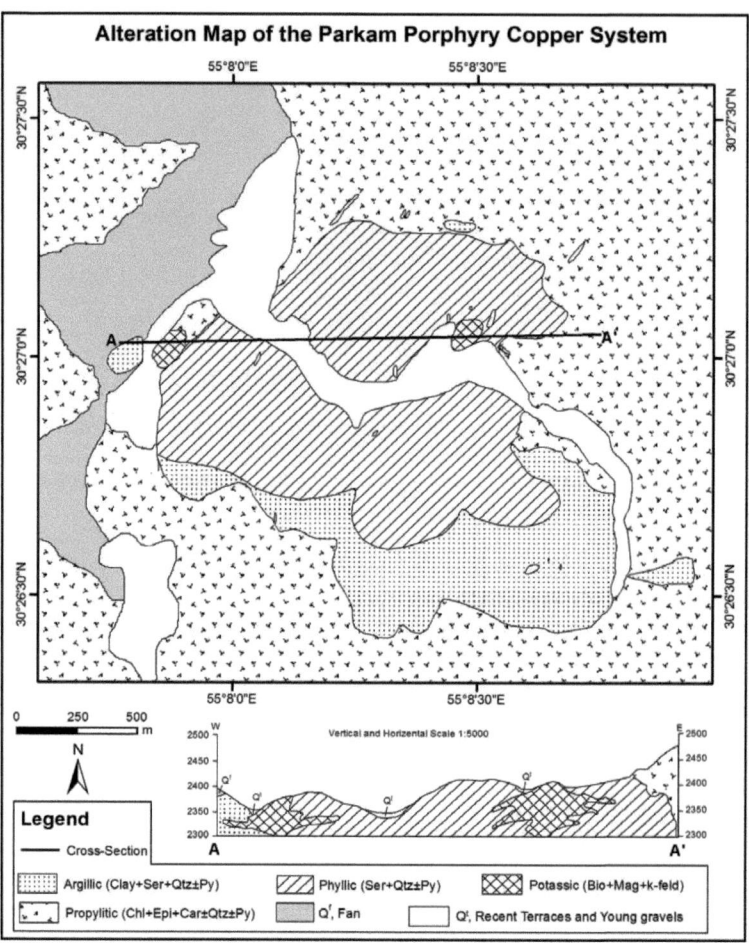

Fig. 6 Alteration map of the Parkam exploration district prepared based on field investigations and petrographical data, and a cross-section describing the vertical and lateral distribution of different alterations (Ser: sericite; Qtz: quartz; Py: pyrite; Chl: chlorite; Epi: epidote; Bio: biotite; Mag; magnetite; K-feld: potassium feldspar; Car: carbonates; Clay: clay minerals).

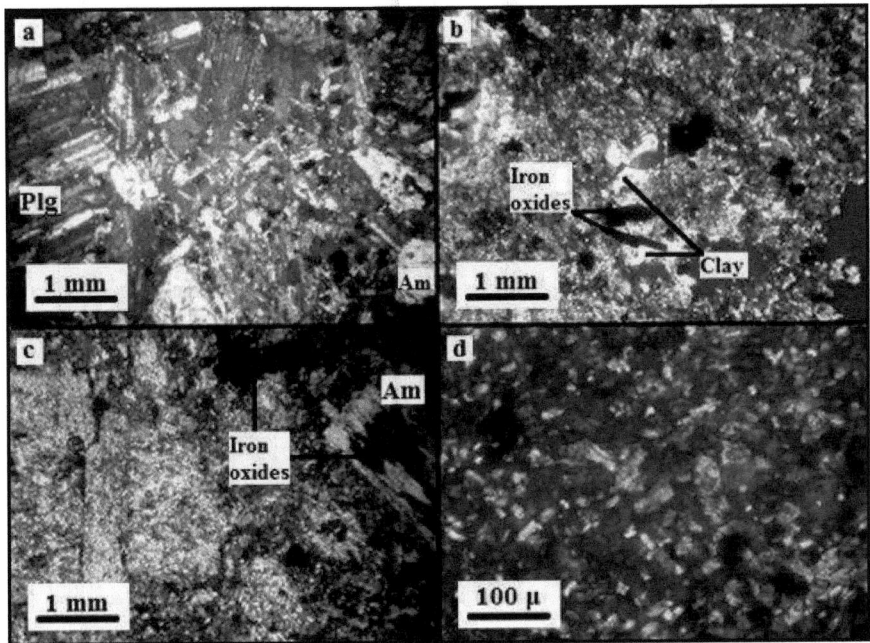

Fig. 7 Photomicrographs of the dioritic subvolcanic rocks in thin section (XPL). a) Diorite with unaltered plagioclase and amphibole. The sericites indicate moderate phyllic alteration; b) Clay minerals associated with iron oxides in a sericitized matrix indicating phyllic and argillic alterations in the subvolcanic rocks; c) Amphibole crystal altered to iron oxides and chlorite, and intense sericitization in the groundmass representing propylitic and phyllic alterations in the dioritic rocks respectively; d) Secondary biotite indicating potassic alteration in the dioritic subvolcanic body (Plg: plagioclase; Am: amphibole).

Mineralization occurs predominantly as fracture-filling hydrous copper carbonates with subordinate amounts of chrysocolla and primary sulfides at the outcrops of the Parkam porphyry system, and is mostly associated with the potassic alteration and its adjacent zone of phyllic alteration; however, malachite may occasionally be observed in other parts of the alteration system including in the fractures of the pyroclastic rocks of the propylitic alteration zone (Fig. 4; Fig. 6; Fig. 8) (Ghannadpour et al. 2015c).

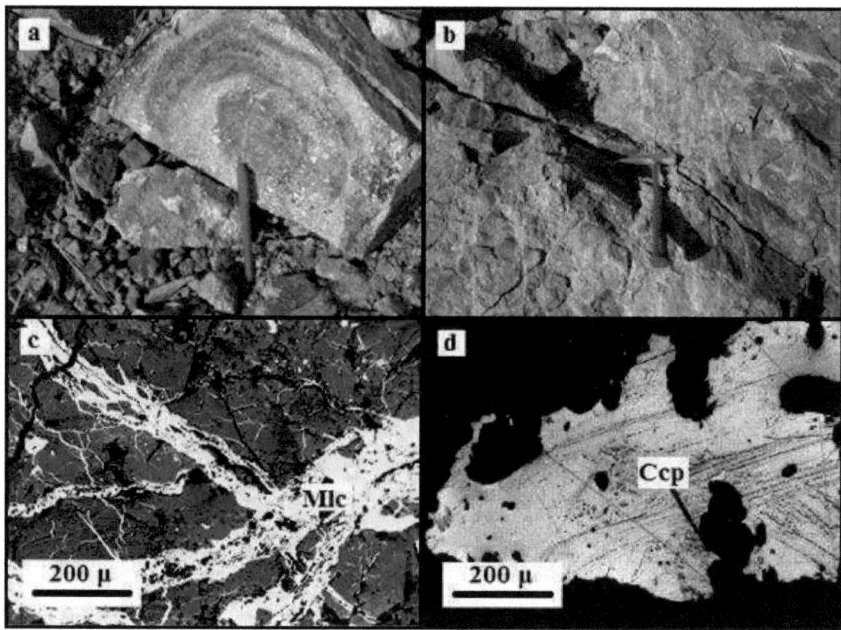

Fig. 8 Surface mineralization of copper in the Parkam porphyry system. a) Malachite in the fractures of the diorite dykes of the potassic zone; b) Malachite in the phyllic zone of the dioritic subvolcanic body; c) Photomicrograph (reflected light) of malachite veinlets filling the fractures of the dioritic rocks of the potassic zone; d) Photomicrograph (reflected light) of a decomposing crystal of chalcopyrite from the potassic zone with its rims dissolved during supergene processes (potassic zone; Mlc: malachite; Ccp: Chalcopyrite).

Ghannadpour et al. (2015b) has compared Parkam deposit with other similar ones like Miduk, Sarcheshmeh and Sungun deposits in Iran from different aspects (Table 1). Also this deposit has been compared with famous deposits of the world in Table 2 (Ghannadpour et al. 2015b).

Table 1 Comparing Parkam deposit with Miduk, Sarcheshmeh and Sungun deposits, Iran.

Comparative Criterion	Parkam Deposit	Miduk Deposit	Sarcheshmeh Deposit	Sungun Deposit
Tectonic Position	Continental Margin	Continental Margin	Continental Margin	Continental Margin
Intrusive Mass	Diorite, Quartz Diorite[9]	Diorite, Quartz Diorite, Granodiorite[5]	Granodiorite, Quartz Monzonite[3]	Monzonite-Quartz Monzonite, Diorite-Granodiorite, Quartz Monzonite[1]
Host Rock	Andesite to Trachyandesite, Pyroclastic[9]	Andesite-Basalt[5]	Andesite[3]	Dacite, Trachyandesite[2]
Alteration	Potassic, Biotite, Potassic-Phyllic, Phyllic, Argillic, Propylitic[9]	Magnetite rich Potassic, Potassic, Potassic-Phyllic, Phyllic, Argillic, Propylitic[5]	Potassic, Propylitic, Phyllic ± Argillic[3]	Potassic, Propylitic, Albite, Potassic-Phyllic, Phyllic, Argillic[2]
Anhydrite Presence	Extensive	Extensive	Numerous	Numerous
Deposit Reserve (million ton)	-	170[5]	1200[3]	600[2]
Cu Average Grade (%)	0.16[9]	0.82[5]	0.7[3]	0.76[2]
Mo Average Grade (%)	0.001[9]	0.007[5]	0.03[3,4]	0.01[2]
Occurrence of Enriched Part of Sulfide Deposit	Weak	Very Extensive	Very Extensive	Limited
Age	13.3 ± 1.1 Ma (^{40}Ar-^{39}Ar)[8]	12.5 ± 0.1 Ma (U-Pb)[6] 10.8 ± 0.4 Ma (^{40}Ar-^{39}Ar)[8] 12.23 ± 0.07 Ma (Re-Os)[5]	13.6 ± 0.1 Ma (U-Pb)[6] 12.1 ± 0.6 Ma (^{40}K-^{40}Ar)[7]	-
Valuable Metal Components	Cu	Cu, Au, Ag[5]	Cu, Mo, Au, Ag[4]	Cu, Mo, Ag[2]

[1]Hezarkhani, 2006b; [2]Hezarkhani and William-Joins, 1998; [3]Samani, 1998; [4]Shahabpour, 2000; [5]Taghipour, et al. 2008; [6]McInnes, et al. 2005; [7]Ghorashizadeh, 1978; [8]Hassanzadeh, 1993; [9]Mohammadi-Laghab, et al. 2012.

Table 2 Comparing Parkam deposit with other porphyry copper deposits of the world.

Comparative Criterion	Parkam Deposit	Northern Chile Deposits		Philippines Deposits
		Cu-Au	Cu-Mo	
Tectonic Position	Continental Margin[3]	Continental Margin[1]	Continental Margin[1]	Arc Islands[1]
Intrusive Mass	Diorite, Quartz Diorite[3]	Diorite (Quartz Diorite)[1]	Tonalite, Granodiorite, Monzogranite[1]	Diorite, Quartz Diorite[1]
Volcanic Rock	Andesite, Tuff[3]	Andesite, Dacite[1]	Andesite[1]	Andesite, Dacite[1]
Potassic Alteration	Related to ore, relatively numerous alkaline feldspar[3]	Related to ore, low alkaline feldspar[1]	Related to Ore, numerous alkaline feldspar[1]	Related to ore, low alkaline feldspar[1]
Phyllic Alteration	Extensive[3]	Limited[1]	Extensive[1]	Limited[1]
Primary Oxidized Fe Mineral	Low and particularly existing in hydrothermal breccia	Numerous[1]	Can be seen rarely[1]	Numerous[1]
Tonnage (million ton)	-	100-150	1500	150-250
Hydrothermal Breccia	Present[3]	Common[1]	Common[1]	Common[1]
Cu Grade (%)	0.16	0.26-1.08[2]	0.3-1.58[2]	0.27-0.8[2]
Mo Grade (%)	0.001	0.006-0.011	0.005-0.04	0.0005-0.022
Secondary Sulfide Enrichment	Weak[3]	Absent[1]	Very Extensive[1]	Absent[1]

[1]Vila and Sillitoe, 1991; [2]Singer, et al. 2008; [3]Mohammadi-Laghab, et al. 2012.

5- Material

National Iranian Copper Industries Co (NICICO) has performed drilling of 7 exploratory boreholes in Parkam area. Coring and chemical analysis produced elemental concentration data as Zn, Pb and Mo in ppm and Fe and Cu in weight percent. The samples have been collected by researchers in NICICO and were analyzed for Cu and Mo by X-ray Fluorescence Spectroscopy (Model: PW-2404, Producer Company: Philips) at the Central Laboratory of the Sarcheshmeh Copper Complex, and in the Chemistry Laboratory of Earth Sciences Development Company, Tehran, Iran. Characteristics of boreholes, rock type and texture in the different depth of boreholes for 1261 samples are shown in Table 3 and the borehole figures (their view from the top and the side) are also depicted in Fig. 9.

Table 3 Characteristics of boreholes, rock type and texture in the different depth of boreholes (QMZ: Quartz Monzonite, DIO: Diorite, ANS: Andesite, POR: Porphyry, QDI: Quartz Diorite, GRL: Granular, GRD: Granodiorite, APH: Aphanitic and TUF: Tuff).

Borehole No.	X COLLAR (m) Y COLLAR (m) Z COLLAR (m)	Depth (m) Dip (deg) Strike	From – To (m)	Rock - Type	Texture	From - To (m)	Rock - Type	Texture
1	321109 3370282 2406	366 70 N80E	0 – 72 72 - 164 164 - 187	QMZ QMZ DIO	POR POR POR	187 - 203 203 - 350 350 - 366	ANS DIO DIO	POR POR POR
2	321520 3370188 2422	592 70 N30E	0 - 104 104 - 236 236 - 312	ANS DIO ANS	GRL POR GRL	312 - 372 372 - 534 534 - 592	DIO QDI ANS	GRL POR POR
3	321389 3370183 2415	597 70 N45W	0 - 50 50 - 77 77 - 220	ANS GRD ANS	POR POR POR	220 - 236 236 - 410 410 - 597	DIO ANS DIO	APH POR POR
4	321447 3370275 2414	347 70 N30W	0 - 46 46 - 77 77 - 143	ANS GRD ANS	POR POR POR	143 - 166 166 - 300 300 - 347	DIO ANS DIO	GRL APH POR
5	321126 3369853 2418	560 70 S60E	0 - 186 186 – 298 298 - 380	ANS ANS ANS	POR POR POR	380 - 415 415 - 534 534 - 560	DIO QDI DIO	POR GRL POR
6	320973 3369992 2390	600 70 N55W	0 - 338 338 - 340 340 - 370	DIO ANS DIO	POR GRL POR	370 - 383 383 - 514 514 - 600	ANS DIO ANS	POR GRL POR
7	321597 3369676 2358	353 70 S45W	0 - 140 140 - 152 152 - 178	DIO QDI DIO	APH POR POR	178 - 266 266 - 272 272 - 353	TUF ANS TUF	APH POR APH

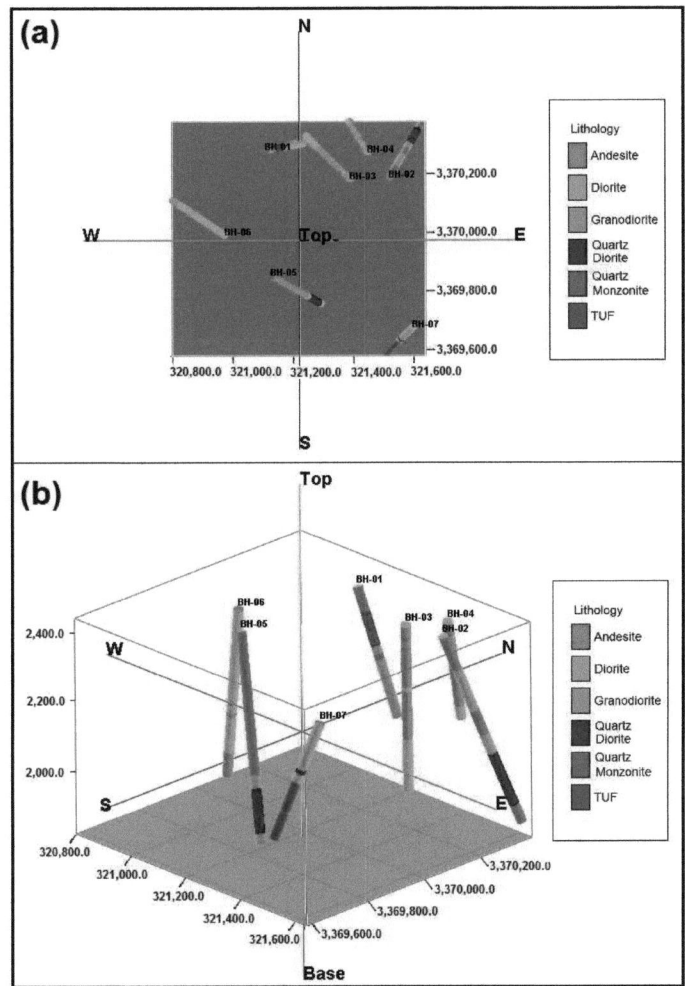

Fig. 9 Figures of boreholes. a) Top view; b) Side view.

Moreover, regular sampling grid with square cells having a side of 100 meters has been executed in the Parkam exploration district. There were 377 samples collected and analyzed for major and trace element concentration by ICP-OES

23

and ICP-MS at Amdel Mineral Laboratories, Adelaide, South Australia. Fig. 10 shows the location of these samples as well as the lithology from which they were collected. The samples located in the fields of the sediments (Fig. 10) were collected either from rock outcrops or from trenches having access to rock, and mostly shared the lithology of their adjacent igneous units. There are several locations with no sample in the grid (Fig. 10). There was no access to rock at these locations due either to the absence of outcrops or the relatively great thickness of the sediments or soil that obstructed manual digging.

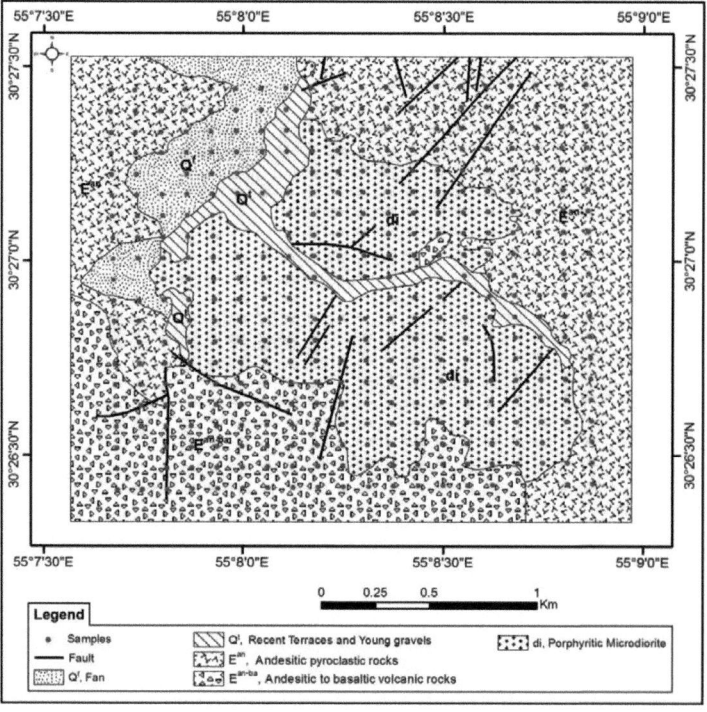

Fig. 10 Magnified part of the geologic map in Figure 4 displaying the sampling locations.

6- Results and Discussions

In this section, at the first step, the Cu, Mo, Pb and Zn analysis and their relative geochemical characteristics will be evaluated in the case of borehole data set. Then, Cu grade will be estimated considering the latitude, longitude and Mo grade of samples also in the case of surface data set.

6-1- The study of the borehole data set in Parkam

In every meter of drilling in each borehole (considered as a sample vector (x) in the drilling collection of NICICO including several tens of thousands of meters) is considered as a vector with characteristics as Zn, Pb, Mo, Fe and Cu grade values, analysis based on each of which might provide suitable perspective for decision making. For instance, some classifications beneficial for geochemical explorations could be as follows:

a) Copper variations relative to Zn.

b) Copper variations relative to Mo.

c) Molybdenum variations relative to Fe.

d) Copper variations relative to Pb.

e) Etc.

In this section, for instance, Cu grade variations relative to Mo, Pb and Zn which are from very important elements in determining extent and spread of primary geochemical halos is considered for clustering in the way of performing the behavioral analysis.

In the present research, the appropriate number of classes (K) was selected by ranging K from 3 to 10 and analyzing the derived classifications. For evaluating the groups resulted from different values of K, the following equation for examining the classifications has been applied (Ghannadpour and Hezarkhani, 2015b):

$$(6) \qquad S(i) = \frac{Min\big(MEAN\_BETWEEN(i,k)\big) - MEAN\_WITHIN(i)}{Max[MEAN\_WITHIN(i), Min\big(MEAN\_BETWEEN(i,k)\big)]}$$

The parameter S(i) in the equation (6) represents the suitability of the ith indexed specimen in its associated class, the parameter "MEAN_WITHIN(i)" stands for the average distance between the ith specimen and the rest of specimens in a particular class and the parameter "MEAN_BETWEEN(i,k)" represents the average distance between the ith specimen and specimens present in an alternative class as K (Ghannadpour and Hezarkhani, 2015b).

According to the equation (6), suitability could range between -1 (inappropriate classification) and +1 (appropriate classification). Also 0 implies indifference for the specimen being classified either in associated class or else.

6-1-1 Geochemical behavior analysis of Cu relative to Mo

Figure 11 represents the silhouette value that indicates the suitability of each specimen of Cu and Pb case testing with 3 and 4 classes (K).

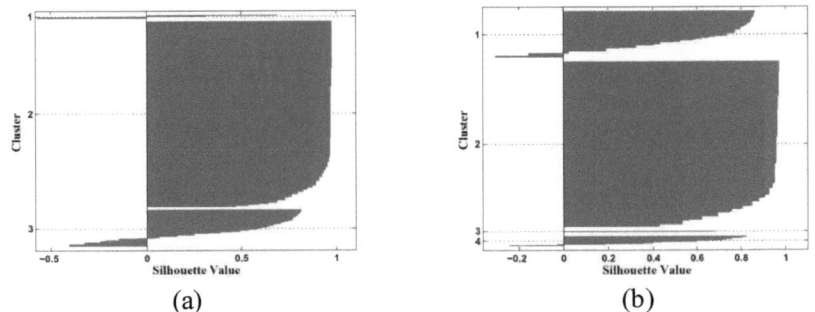

(a)   (b)

Fig. 11 Classes silhouette and suitability value for classification relating to Cu and Mo. (a) Classification with K=3, average value of 0.8437; (b) Classification with K=4, average value of 0.8281.

As represented, the classes 1 and 2 have appropriate silhouette values for K=3 and there are nearly few negative silhouette values associated with class 2. Additionally, the average silhouette value is 0.8437. There are also negative silhouette values in K=4 testing (having an average silhouette value of 0.8218). Obviously, clustering with K=3 is preferred to K=4 (due to the greater value of average silhouette). The same is performed for K=5 to K=10 cases (Fig. 12).

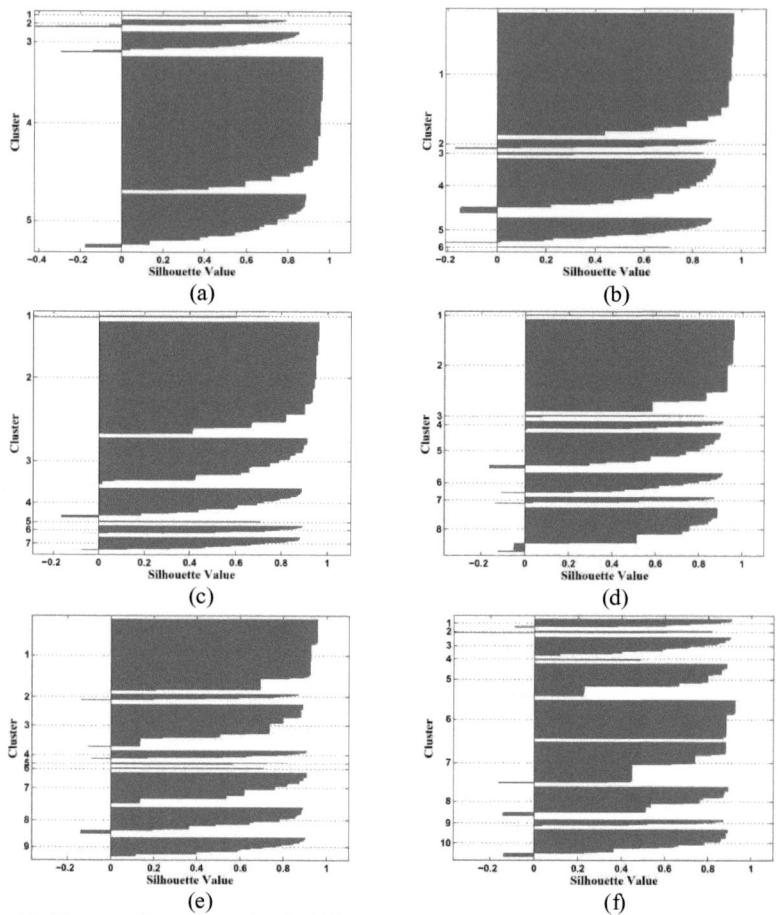

Fig. 12 Classes silhouette and suitability value for classification relating to Cu and Mo. (a) Classification with K=5, average value of 0.7847; (b) Classification with K=6, average value of 0.7926; (c) Classification with K=7, average value of 0.7808; (d) Classification with K=8, average value of 0.7483; (e) Classification with K=9, average value of 0.7294; (f) Classification with K=10, average value of 0.5881.

Eventually, considering the results for K=3 to K=10, clustering with K=3 for the specimens of Cu and Mo grade attributes was selected. The central points of the K=3 classification are represented in Fig. 13. As illustrated, Mo grade varies linearly relative to Cu grade values.

The best regression was $y = 15553x - 221.5$ and the correlation coefficient was reported $R^2 = 0.8942$. Figure 13 represents the regression line fitted to class central points of Cu and Mo case as well.

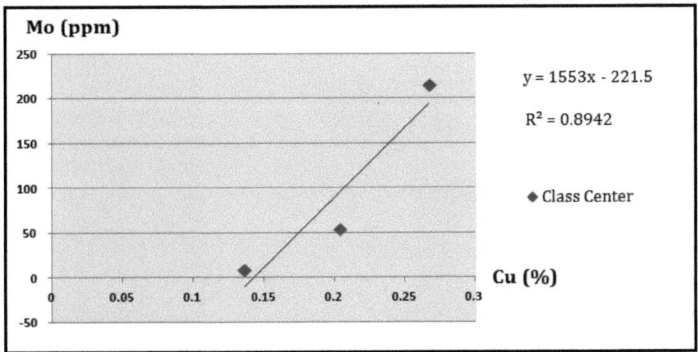

Fig. 13 The central points of the K=3 classification relating to Cu and Mo, and regression line crossed from them.

6-1-2 Geochemical behavior analysis of Cu relative to Pb

Figure 14 represents the silhouette value which indicates the suitability of each specimen presented in classification with K=3 to K=10 in the case of Cu and Pb. The considering the result derived from analysis with K=3 to K=10, as illustrated, the classification with K=4 for specimens having characteristics of Cu and Pb grade values best satisfies the criteria of data mining suggested above (see equation 6). The central points of classes derived from clustering with K=4 are represented in Fig. 15.

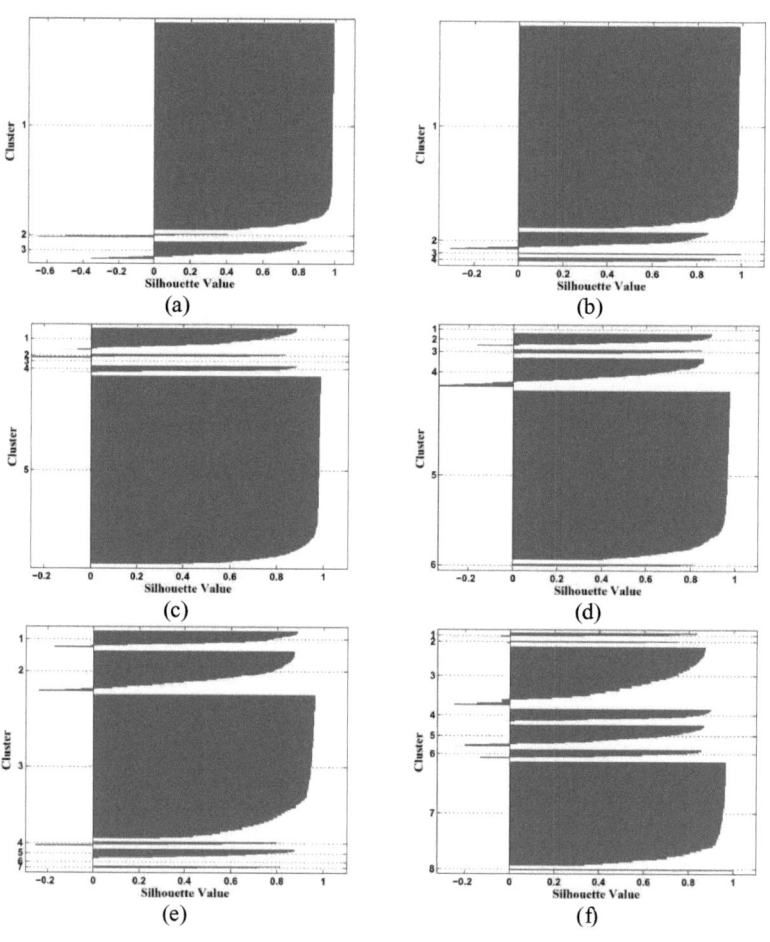

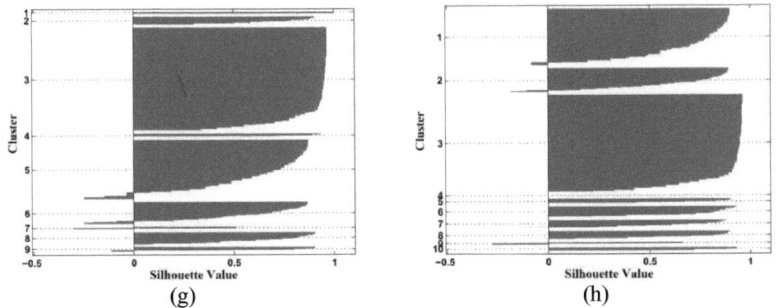

(g)　　　　　　　　　　　　　　　　(h)

Fig. 14 Classes silhouette and suitability value for classification relating to Cu and Pb. (a) Classification with K=3, average value of 0.9161; (b) Classification with K=4, average value of 0.9292; (c) Classification with K=5, average value of 0.9052; (d) Classification with K=6, average value of 0.8469; (e) Classification with K=7, average value of 0.7972; (f) Classification with K=8, average value of 0.7439; (g) Classification with K=9, average value of 0.5737; (h) Classification with K=10, average value of 0.7310.

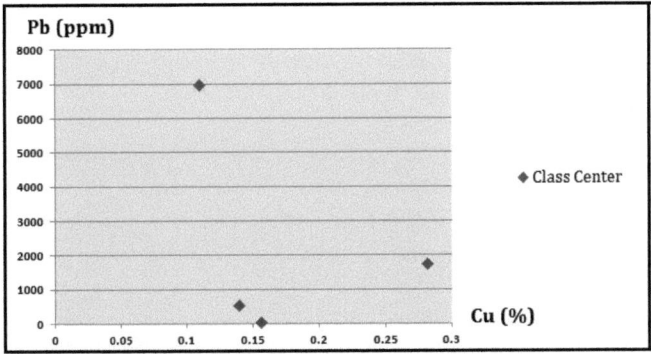

Fig. 15 The central points of the K=4 classification relating to Cu and Pb.

Growth in Cu grade values results in an initial abrupt decrease in Pb grade values and after greater Cu grade values than 0.16, Pb grade values increase with a gentle slope. Considering this fluctuation, the best curve fitting the central points was $y = 10^6 x^2 - 431877x + 41741$ and the correlation coefficient was reported $R^2 = 0.951$.

6-1-3 Geochemical behavior analysis of Cu relative to Zn

The suitability of each specimen for classifying with K=3 to K=10 in the case of Cu and Zn is represented in Fig. 16.

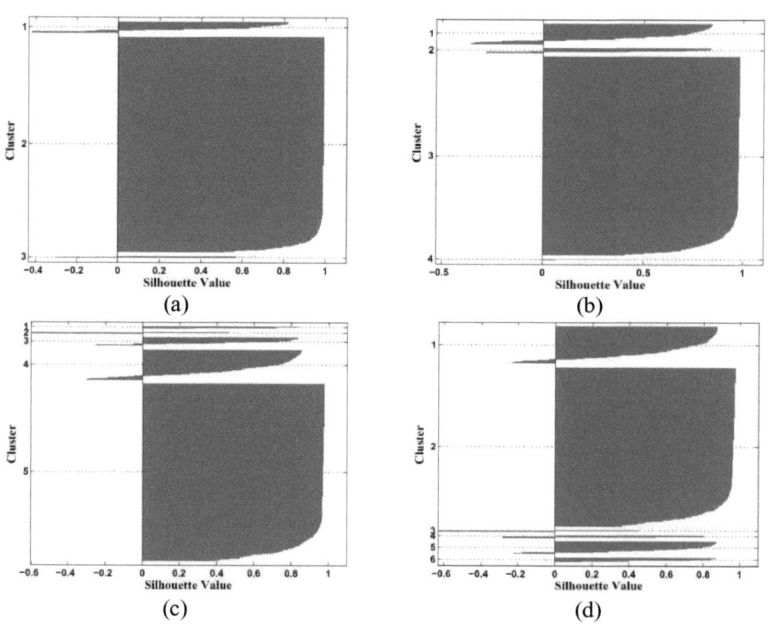

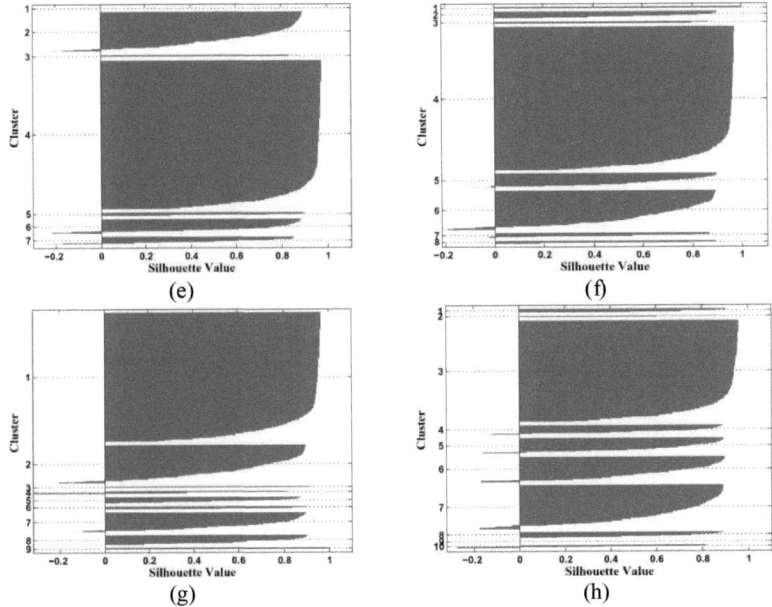

Fig. 16 Classes silhouette and suitability value for classification relating to Cu and Zn. (a) Classification with K=3, average value of 0.9439; (b) Classification with K=4, average value of 0.8985; (c) Classification with K=5, average value of 0.8587; (d) Classification with K=6, average value of 0.8385; (e) Classification with K=7, average value of 0.8296; (f) Classification with K=8, average value of 0.8293; (g) Classification with K=9, average value of 0.8033; (h) Classification with K=10, average value of 0.7640.

Considering the result derived from the analysis with K=3 to K=10, as illustrated, the classification with K=3 for specimens having characteristics of Cu and Zn grade values best satisfies the criteria of data mining suggested above (see equation 6). The central points of classes derived from clustering with K=3 are represented in Fig. 17.

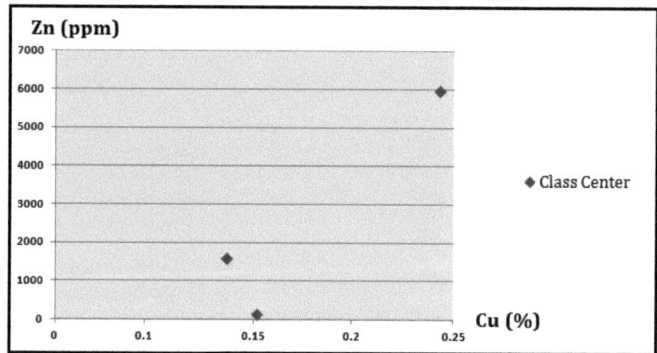

Fig. 17. The central points of the K=3 classification relating to Cu and Zn

Comparable to the Pb case, increase in Cu grade values, causes a decrease and a subsequent increase in Zn grade values. However, the fluctuation is not identical to the case of Pb which is more distinct. In the Pb case there were 4 classes being interpreted and the fluctuations were more distinguishable while in the latter case involving 3 classes the central point's show erratic pattern and the fluctuation must be indicated with caution. However, due to the relationship between Lead and Zinc in Study of Ghannadpour et al (2013) could be realized the zinc increasing and decreasing. The curve fitted on central points was $y = 2 \times 10^6 x^2 - 761065x + 67355$ and the correlation coefficient was reported as $R^2 = 0.973$.

With regards to the decline in the amount of Pb and Zn in a specific range and their subsequent increase with increasing amounts of Cu in the same range it could be concluded that Pb and Zn have been mobilized from the potassic alteration zone to the zone of phyllic alteration. Considering the mineralization in the porphyry systems, if Pb and Zn are mineralized into the potassic zone which is of high temperature and when the meteoric waters are introduced into the system and are becoming trapped as fluid inclusions, stable complexes of Pb and Zn could form and significant amounts of Pb and Zn are leached from the potassic zone and concentrated into the phyllic zone. However, this is not true of the Cu and Mo case. During interaction of meteoric waters with the potassic zone, Cu is only partially mobilized into the phyllic zone and Mo is not mobilized at all. Therefore, sometimes in a Cu porphyry system one hump and/or one minimization of Pb and Zn in relation with Cu is observed and this is due to the Pb and Zn being leached from the potassic zone completely and concentrated into the phyllic zone. Yet this is not the case for the Cu and Mo. Therefore, this change in concentration is associated with different alteration characterizations.

6-2- The study of the surface data set in Parkam

In this section, the algorithm of K-means method is applied on the surface samples of Parkam district in order to investigate the Cu and Mo geochemical behavior, considering the latitude and longitude of samples. Actually, Cu grade is estimated based on three parameters (latitude, longitude and Mo grade of samples). For this purpose, at the first step, each sample is considered as a vector with characteristics as Cu and Mo grade values and also the sample latitude and longitude. Then, algorithm of K-means method is applied on the surface data set.

The silhouette value for classifying with K=3 to K=10 are illustrated in Fig. 18 in the case of Cu and Mo considering the sample latitude and longitude.

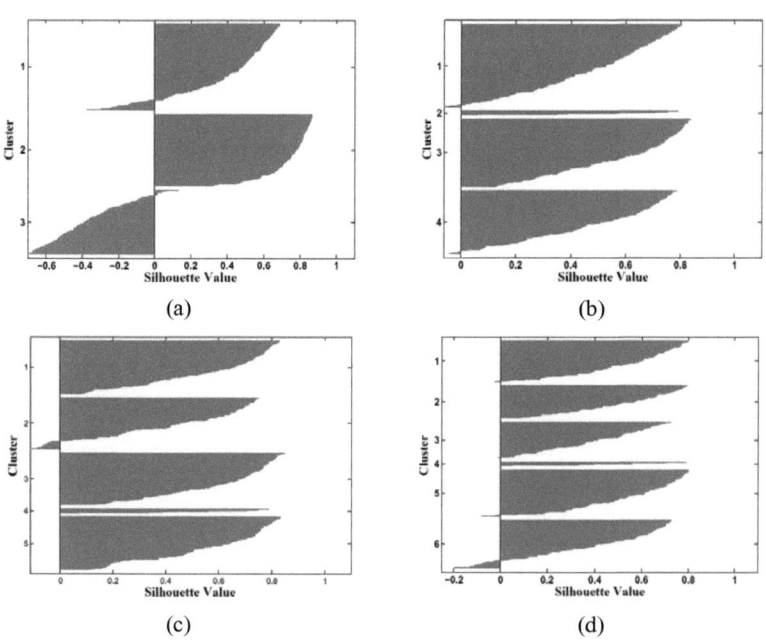

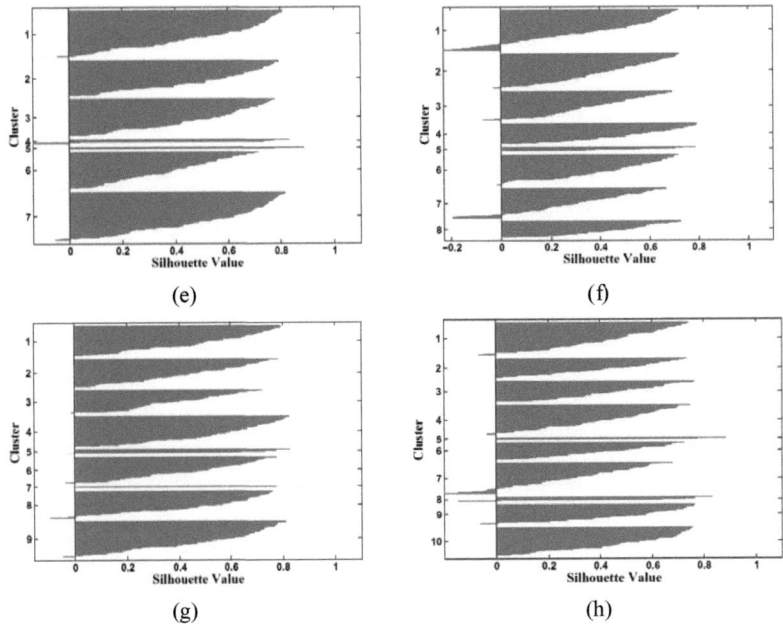

(e)   (f)

(g)   (h)

Fig. 18. Silhouette classes and value suitability for relating classification in Cu and Mo considering their latitude and longitude. (a) Classification with K=3, average value of 0.2828; (b) Classification with K=4, average value of 0.5055; (c) Classification with K=5, average value of 0.5346; (d) Classification with K=6, average value of 0.4711; (e) Classification with K=7, average value of 0.4913; (f) Classification with K=8, average value of 0.3979; (g) Classification with K=9, average value of 0.4824; (h) Classification with K=10, average value of 0.4271.

According to the results for K=3 to K=10, clustering with K=5 for the specimens of Cu and Mo grade based on their latitude and longitude attributes are selected. In the following, an equation will be presented in order to estimate the Cu grade using the five class center values.

6-2-1 Copper and Mo Relationship determination

In this section, by means of the multivariate regression and SPSS software concept, the relationship between the copper and molybdenum will be determined considering the sample latitude and longitude, within Parkam district.

For this purpose, class centers values in the case of Cu grade as the dependent variable and class centers values in the case of Mo grade, latitude and longitude as the independent variables are introduced into the software. Eventually, the results in Table 4 as characteristics and multivariate regression coefficients have been calculated and reported by the mentioned software.

Table 4 Characteristics and multivariate regression coefficients.

R	r^2	A	b_1	b_2	b_3
0.865	0.748	21221585.55	5.072	-6.771	-6524.796

According to the above table, equation of Cu grade estimation is considered as follow:

$$Cu = 21221585.55 + 5.072X - 6.771Y - 6524.796Mo$$

According to the correlation coefficient (r^2), it could be said that, 75% of copper changes through independent parameters such as latitude, longitude and Mo grade of samples have been obtain, approximately. In the other words, Almost 25% in copper changes is not achieved by regression.

It is noteworthy that the mentioned correlation coefficient is related to the plate which has been passed through the cluster centers.

7- Conclusions

In this research, the K-means clustering method was applied for classifying the geochemical data set derived from exploration drillings in the uneconomic Parkam porphyry system. According to the linear relationship between copper and molybdenum and also decreasing and increasing of Pb and Zn, respectively, with increase of copper and molybdenum, we conclude that the action of meteoric waters has led to leaching of the Pb and Zn from the potassic zone and deposition in the phyllic zone, but it did not have the ability to mobilize the Cu and Mo (maybe due to the lower temperatures that could stop the Cu and Mo complexes from mobilizing). It is probable that this is the reason for the low Cu grades in the Parkam porphyry system which makes it uneconomic. It means that there was no chance for Cu and or Mo to be more concentrated in the next maturity processes (phyllic zone) of the Parkam system, as it happens in all economic porphyry Cu-Mo deposits. But, the fluctuations in Pb and Zn within the system (evident in the text), could be the fingerprint of hydrothermal regime changes by Pb and Zn increase and decrease during the evolution of the Parkam system. It is also concluded that the temperature could have a very strong effect on the Parkam maturation as a porphyry system. This characteristic could be important in separating the economic porphyry systems from those of uneconomic ones. Finally, equation of Cu grade estimation according to the mentioned the five cluster centers are presented using K-Means method in the case of surface data set. It is also suggested that for the further studies, a curved plate (corrugated plates) that could be fitted to the center of clusters instead of a plate, despite the complexity of its computations, could be done, since both accuracy and the correlation coefficient would be significantly increased.

References

1. Adamia S, Bergougnan H, Fourquin C, Haghipour A, Lordkipanidze M, Ozgül N, Ricou L, Zakariadze G (1980) The alpine Middle East between the Aegean and the Oman traverses. Aubouin, J.; Debelmas, J.; Latreille, M. (coords.) Geologie des chaines Alpines issues de la Tethys. Bureau de Recherches Géologique et Minières Mémoir no. 116, Orleans, France 122-136

2. Alavi M (1994) Tectonics of the Zagros orogenic belt of Iran: new data and interpretations. Journal of Tectonophysics 229 (3): 211-238

3. Anderberg M R (1973) Cluster analysis for applications. Academic Press, New York

4. Berberian M (1983) The southern Caspian: a compressional depression floored by a trapped, modified oceanic crust. Canadian Journal of Earth Sciences 20 (2): 163-183

5. Berberian M, King G (1981) Towards a pale geography and tectonic evolution of Iran. Canadian Journal of Earth Sciences 18 (2): 210-265

6. Bock H H (2008) Origins and extensions of the k-means algorithm in cluster analysis. Electronic Journal for History of Probability and Statistics 4(2): 1-18

7. Cheung Y M (2003) K-means: A new generalized k-means clustering algorithm. Pattern Recognition Letters 24(15): 2883-2893

8. Devijver P A, Kittler J (1982) Pattern recognition: A statistical approach. Prentice/Hall International Englewood Cliffs, NJ

9. Etminan H (1978) Fluid Inclusion Studies of the Porphyry Copper Ore Bodies at Sar-Cheshmeh, Darreh Zar, and Mieduk (Kerman Region, Southeastern Iran) and Porphyry Copper Discoveries at Miduk, Gozan, and Kighal, Azarbaijan Region (Northwestern Iran) [abs.]: International Association of the Genesis of Ore Deposits. 5th Symposium. Snowbird. Utah. Abstract Volume, p: 88

10. Gent M, Menendez M, Toraño J, Susana T (2011) A review of indicator minerals and sample processing methods for geochemical exploration. Journal of Geochemical Exploration 110: 47-60

11. Ghannadpour S S, Hezarkhani A (2012) Determine the initial statistical specifications of Copper and molybdenum elements in Porphyry Copper ore deposit in Kerman. International Mining Congress and Exploration, Iran

12. Ghannadpour S S (2013) Geochemical studies of porphyry copper ore deposit of Parkam. MSc Thesis, Amirkabir University of Technology, Tehran, p: 308

13. Ghannadpour S S, Hezarkhani A, FarahBakhsh E (2013) An Investigation of Pb Geochemical Behavior Respect to Those of Fe and Zn Based on k-Means Clustering Method. Journal of Tethys 1(4): 291-302

14. Ghannadpour S S, Hezarkhani A, Maghsoudi A (2014) Investigation of distribution and exact calculation of statistical parameters of Fe in Parkam-Kerman: 32nd National and The 1st International Geosciences Congress, Iran. Tehran

15. Ghannadpour S S, Hezarkhani A (2015a) Lead and zinc geochemical behavior based on geological characteristics in Parkam Porphyry Copper System, Kerman, Iran. Journal of Centeral of South University 22(11): 4274-4290

16. Ghannadpour S S, Hezarkhani A (2015b) Investigation of Cu, Mo, Pb, and Zn geochemical behavior and geological interpretations for Parkam porphyry copper system, Kerman, Iran. Arabian Journal of Geosciences 8(9): 7273–7284

17. Ghannadpour S S, Hezarkhani A, Sabetmobarhan A (2015a) Some statistical analyses of Cu and Mo variates and geological interpretations for Parkam Porphyry Copper system, Kerman, Iran. Arabian Journal of Geosciences 8(1): 345–355

18. Ghannadpour S S, Hezarkhani A, Maghsoudi A, Farahbakhsh E (2015b) Assessment of prospective areas for providing the geochemical anomaly maps of lead and zinc in Parkam district, Kerman, Iran. Geosciences Journal 19(3): 431–440

19. Ghannadpour S S, Hezarkhani A, Sabetmobarhan A (2015c) The Parkam exploration district (Kerman, Iran): Geology, alterations, and delineation of Cu- and Mo-mineralized zones using U-spatial statistic with associated software development. Journal of Earth Scieces, DOI: 10.1007/s12583-015-0644-6

20. Ghorashizadeh M (1978) Development of hypogene and supergene alteration and copper mineralization patterns, Sarcheshmeh porphyry copper deposit, Iran. MSc Thesis, Brock University, Canada, p: 223

21. Hallam A (1976) Geology and plate tectonics interpretation of the sediments of the Mesozoic radiolarite-ophiolite complex in the Neyriz region, southern Iran. Geological Society of America Bulletin 87(1): 47-52

22. Hassanzadeh J (1993) Metallogenic and tectono-magmatic events in SE sector of the Cenozoic active continental margin of Central Iran (Shahr-Babak, Kerman province). Ph.D. Thesis, University of California, Los Angeles, p: 201

23. Hezarkhani A, Williams-Jones A E (1998) Controls of alteration and mineralization in the Sungun porphyry copper deposit, Iran: Evidence from fluid inclusion and stable isotopes. Economic Geology 93(5): 651-670

24. Hezarkhani A (2006a) Hydrothermal evolution of the Sar-Cheshmeh porphyry Cu–Mo deposit, Iran: evidence from fluid inclusions. Journal of Asian Earth Sciences 28(4): 409-422

25. Hezarkhani A (2006b) Petrology of the intrusive rocks within the Sungun porphyry copper deposit, Azerbaijan, Iran. Journal of Asian Earth Sciences 27(3): 326-340

26. Hezarkhani A (2008) Hydrothermal Evolution of the Miduk Porphyry Copper System, Kerman, Iran: A Fluid Inclusion Investigation. International Geology Review 50: 1-20

27. Hezarkhani A, Ghannadpour S S (2015) Exploration Information Analysis, first ed. Amirkabir University of Technology (Tehran Polytechnic) press, Tehran, p: 245 (In Persian)

28. Jain A K (2010) Data clustering: 50 years beyond k-means. Pattern Recognition Letters 31(8): 651-666

29. Jébrak M (2006) Economic Geology: Then and Now. Journal of Geoscience Canada 33(2): 81-93

30. K.I.e.c. engineers (Kan Iran exploration consulting engineers) (2009) Geological and Alteration Studies in Parkam Area, Scale: 1/5000

31. Krishna K, Narasimha Murty M (1999) Genetic k-means algorithm. Systems, Man, and Cybernetics, Part B: Cybernetics, IEEE Transactions on 29(3): 433-439

32. Lawrence S, Giles C L, Tsoi A C (1996) What Size Neural Network Gives Optimal Generalization? Convergence Properties of Back propagation. Technical Report UMIACS-TR-96-22 and CS-TR-3617 Institute for

Advanced Computer Studies University of Maryland College Park, MD 20742

33. McInnes B I A, Evans N J, Fu F Q, Garwin S (2005) Application of thermo chronology to hydrothermal ore deposits. Reviews in Mineralogy and Geochemistry 58(1): 467-498

34. Menard J (1995) Relationship between altered pyroxene diorite and the magnetite mineralization in the Chilean Iron Belt, with emphasis on the El Algarrobo iron deposits (Atacama region, Chile). Mineralium Deposita 30(3-4): 268-274

35. Meshkani S A, Mehrabi B, Yaghubpur A, Fadakar Alghalandis Y (2011) The application of geochemical pattern recognition to regional prospecting: A case study of the Sanandaj–Sirjan metallogenic zone, Iran. Journal of Geochemical Exploration 108: 183-195

36. Mohammadi-Laghab H, Taghipour N, Iranmanesh M R (2012) Dispersion pattern of Cu, Mo and Pb, Zn and Fe in Sarah (Parkam) porphyry copper deposit, Shahr-e Babak, Kerman province, Iran. Iranian Journal of Geology 5(11): 17-28

37. Mora J, Armas-Herrera C, Guerra J (2012) Factors affecting vegetation and soil recovery in the Mediterranean woodland of the Canary Islands (Spain). Journal of Arid Environments 87: 58-66

38. Murthy C A, Chowdhury N (1996) In search of optimal clusters using genetic algorithms. Pattern Recognition Letters 17(8): 825-832

39. Nelson P A, Bellugi D, Dietrich W E (2012) Delineation of river bed-surface patches by clustering high-resolution spatial grain size data. Journal of Geomorphology, 205: 102-119

40. Pelleg D, Moore A (1999) Accelerating exact k-means algorithms with geometric reasoning. Proceedings of the fifth ACM SIGKDD international conference on Knowledge discovery and data mining, ACM: 277-281

41. Saha S, Bandyopadhyay S (2013) A generalized automatic clustering algorithm in a multiobjective framework. Journal of Applied Soft Computing 13(1): 89-108

42. Samani B (1998) Distribution, setting and metallogenesis of copper deposits in Iran. In: Porphyry and hydrothermal copper and gold deposits: a global Perspective. Conference Proceedings 135–158.

43. Saric A, Diordjevic M, Dimitrijevic M N (1971) Geological map of Shahre-e-Babak, 1:100,000 Series. Geological Survey of Iran, Tehran, Iran

44. Sfidari E, Kadkhodaie-Ilkhchi A, Najjari S (2010) Comparison of intelligent and statistical clustering approaches to predicting total organic carbon using intelligent systems. Journal of Petroleum Science and Engineering 86-87: 190-205.

45. Shahabpour J (1982) Aspects of alteration and mineralization at the Sarcheshmeh copper-molybdenum deposit, Kerman, Iran. Ph.D. thesis, University of Leeds, England

46. Shahabpour J (2000) Behavior of Cu and Mo in the Sarcheshmeh porphyry Cu deposit, Kerman, Iran. Canadian Institute of Mining, Metallurgy and Petroleum Bulletin 93: 44–51

47. Shahabpour J (2005) Tectonic evolution of the orogenic belt in the region located between Kerman and Neyriz. Journal of Asian Earth Sciences 24(4): 405-417

48. Singer D A, Berger V I, Moring B C (2008) Porphyry copper deposit in the world: Database, map and grade and tonnage models. (USGS Open-file Report 2008–1060) US Geological Survey, Denver

49. Stöcklin J (1974) Possible ancient continental margins in Iran: The geology of continental margins 873-887

50. Stöcklin J (1977) Structural Correlation of the Alpine Range between Iran and Central Asia. Memoire Hors-Serve No. 8 dela Societe Geologique de France 8: 333-353

51. Taghipour N, Aftabi A, Mathur R (2008) Geology and Re-Os geochronology of mineralization of the Miduk porphyry copper deposit, Iran. Resource Geology 58(2): 143–160

52. Tahmasebi P, Hezarkhani A (2012) A hybrid neural networks-fuzzy logic-genetic algorithm for grade estimation. Computers and Geosciences 42: 18-27

53. Takin M (1972) Iranian geology and continental drift in the Middle East. Journal of Nature 235: 147-150

54. Tarkian M, Stribrny B (1999) Platinum-group elements in porphyry copper deposits: a reconnaissance study. Journal of Mineralogy and petrology 65(3-4): 161-183

55. Vila T, Sillitoe R H (1991) Gold–rich porphyry systems in the Maricunga belt, northern Chile. Economic Geology 86(6): 1238–1260

56. Walker, R., Jackson, J., 2002. Offset and Evolution of the Gowk Fault, S.E, Iran: A Major Intra-Continental Strike-Slip System. Journal of Structural Geology 24: 1677–1698

57. Wegner T, Hussein T, Hämeri K, Vesalad T, Kulmalad M, Weber S (2012) Properties of aerosol signature size distributions in the urban environment as derived by cluster analysis. Atmospheric Environment 61:350-360

58. Welland M J, Mitchell A (1977) Emplacement of the Oman ophiolite: A mechanism related to subduction and collision. Geological Society of America Bulletin 88(8): 1081-1088

59. Xu L, Bi X, Hu R (2012) Relationships between porphyry Cu–Mo mineralization in the Jinshajiang–Red River metallogenic belt and tectonic

activity: Constraints from zircon U– Pb and molybdenite Re– Os geochronology. Ore Geology Reviews 48: 460-473

60. Yaghini M, Ghannadpour S F, Khedmatlou S (2008) Representing an innovative clustering method in data mining using genetic algorithm and solving a real case study in railway transportation industry. Iran Data mining Conference, Amirkabir University of Technology, Iran (In Persian)

61. Yang J, Zhuang Y, Wu F (2012) ESVC-based extraction and segmentation of texture features. Computers and Geosciences 49: 238-247

 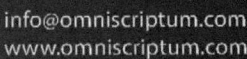

Printed by Books on Demand GmbH, Norderstedt / Germany